Addition and subtraction

This book belongs to

....

Colour the star when you complete a page.
See how far you've come!

Author: Peter Clarke

How to use this book

- Find a quiet, comfortable place to work, away from distractions.

- Ask your child what addition and subtraction topic they are doing at school and choose an appropriate topic.

- Tackle one topic at a time.

- Help with reading the instructions where necessary and ensure that your child understands what they are required to do.

- Help and encourage your child to check their own answers as they complete each activity.

- Discuss with your child what they have learnt.

- Let your child return to their favourite pages once they have been completed, to play the games and talk about the activities.

- Reward your child with plenty of praise and encouragement.

Special features

- Yellow outlined boxes at the top of a page: Introduce and outline the key addition or subtraction ideas.

- Example boxes: Show how to do the activity.

- Yellow shaded boxes at the bottom of a page: Offer advice on how to consolidate your child's understanding.

- Games: Some of the topics include a game, which reinforces the addition or subtraction topic. Some of these games require a spinner. This is easily made using a pencil, a paperclip and the circle(s) printed on each games page. Place the pencil and paperclip at the centre of the circle and then flick the paperclip to see where it lands.

Published by Collins
An imprint of HarperCollins*Publishers*
1 London Bridge Street
London SE1 9GF

HarperCollins*Publishers*
1st Floor, Watermarque Building,
Ringsend Road, Dublin 4, Ireland

First published 2011
This edition © HarperCollins*Publishers* 2022

20 19 18 17

ISBN 978-0-00-813428-0

The author asserts his moral right to be identified as the author of this work.

The author wishes to thank Brian Molyneaux for his valuable contribution to this publication.

All rights reserved. No part of this publication may be reproduced, stored in a retrieval system, or transmitted, in any form or by any means, electronic, mechanical, photocopying, recording or otherwise, without the prior permission of Collins.

British Library Cataloguing in Publication Data.

A Catalogue record for this publication is available from the British Library.

Written by Peter Clarke
Page design by Sarah Duxbury
Cover design by Sarah Duxbury and Amparo Barrera
Project managed by Chantal Peacock, Tracey Cowell and Sonia Dawkins
All images are © HarperCollins*Publishers* Ltd or © Shutterstock.com
Printed in India by Multivista Global Pvt. Ltd

Contents

How to use this book	2
Understanding addition	4
Understanding subtraction	6
Addition facts to 10	8
Subtraction facts to 10	10
Addition facts to 20	12
Subtraction facts to 20	14
Multiples of 10	16
Adding three 1-digit numbers	18
Adding a 2-digit number and ones	20
Subtracting a 2-digit number and ones	22
Adding a 2-digit number and tens	24
Subtracting a 2-digit number and tens	26
Adding two 2-digit numbers	28
Subtracting two 2-digit numbers	30
Answers	32

Understanding addition

Addition (+) is finding the **total** of two or more numbers.

4 + 3 = 7

1 Find the totals.

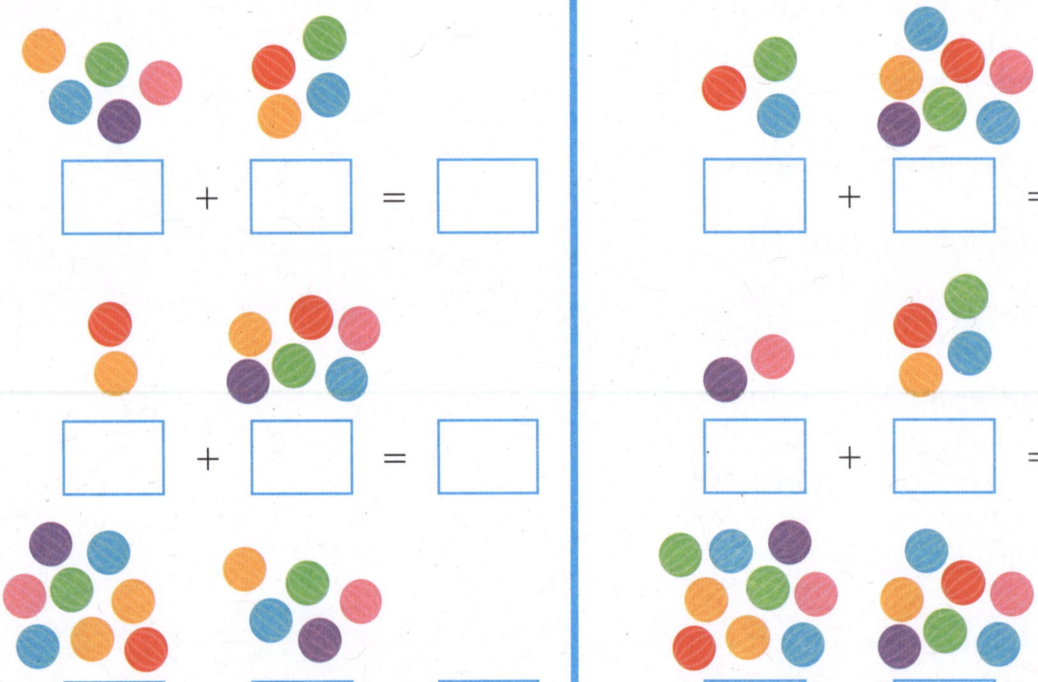

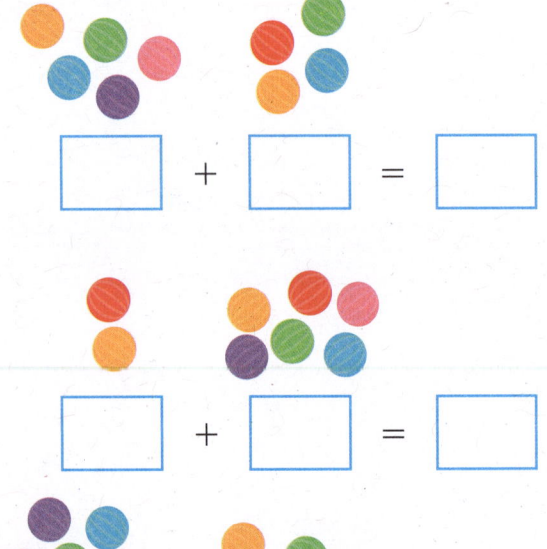

2 Add the same coloured sweets.

A number line helps with addition. Use it to **count on** from the larger number.

3 + 5 = 8

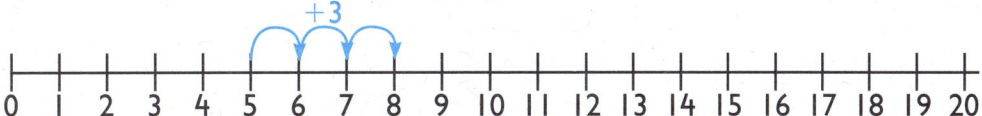

3 Find the totals.

4 + 6 = ☐

8 + 3 = ☐

5 + 7 = ☐

6 + 9 = ☐

12 + 5 = ☐

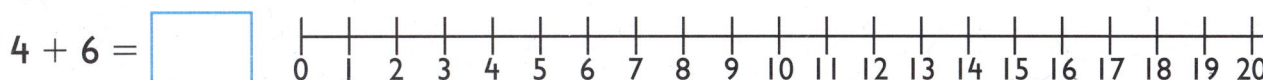

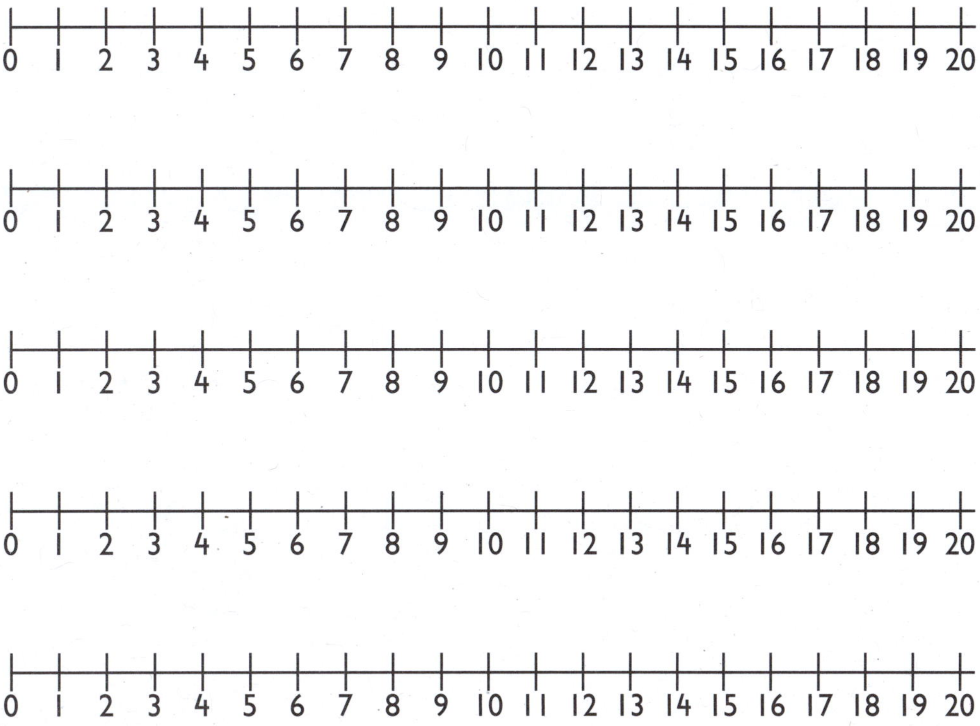

4 Draw lines to join each addition problem with its total.

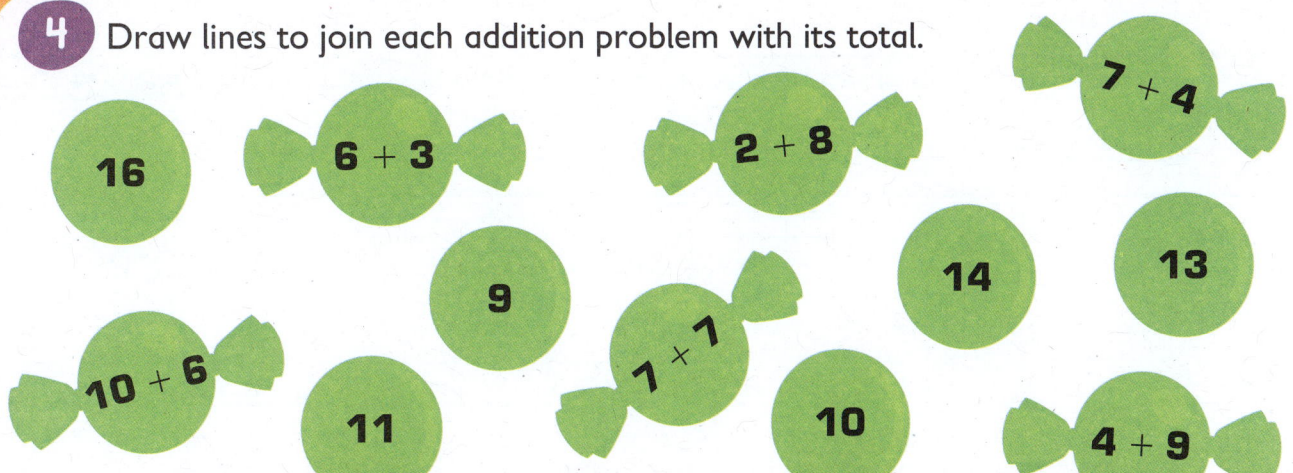

Look for situations where your child can add together two groups of objects, e.g. knives and forks on the table, people in two supermarket queues, apples and oranges in a bowl.

Understanding subtraction

We can think of subtraction (−) as **taking away**.

         7 − 3 = 4

You can **count back** from the larger number along a number line to help you take away.

1 Cross out the fruits and write the answers.

5 − 2 = ☐

6 − 4 = ☐

9 − 5 = ☐

8 − 3 = ☐

12 − 6 = ☐

10 − 2 = ☐

2 Count back from the larger number to help you answer these.

7 − 5 = ☐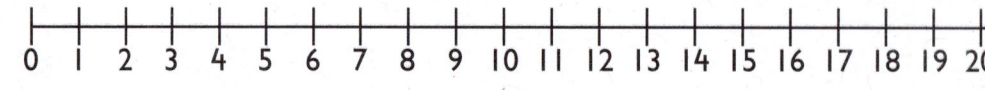

10 − 6 = ☐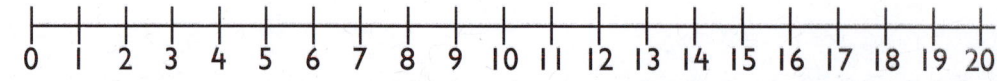

18 − 9 = ☐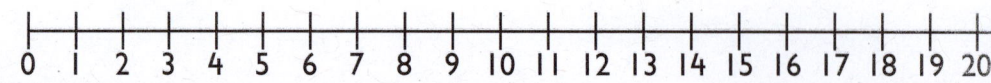

15 − 8 = ☐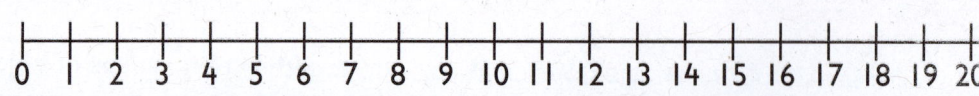

We can also think of subtraction (−) as **finding the difference** between two numbers.

You can **count on** from the smaller number along a number line to help you find the difference.

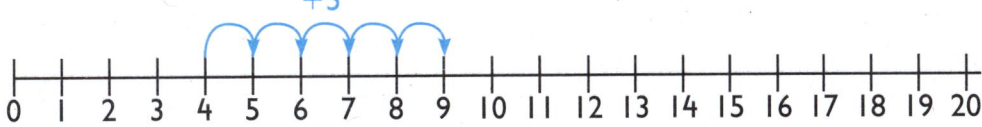

3 Count on from the smaller number to help you answer these.

7 − 2 = ☐

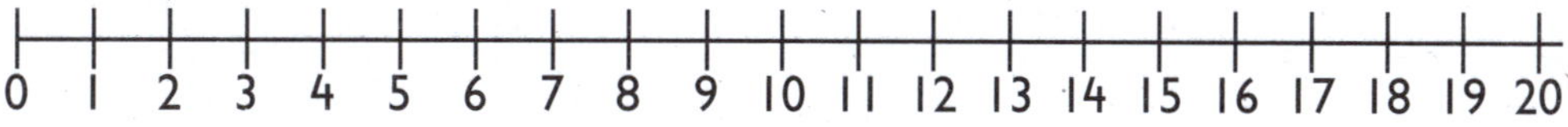

16 − 9 = ☐

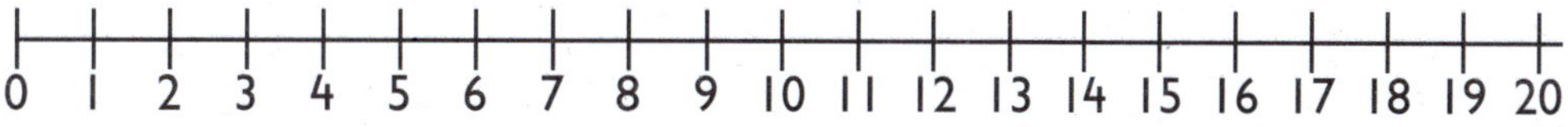

12 − 5 = ☐

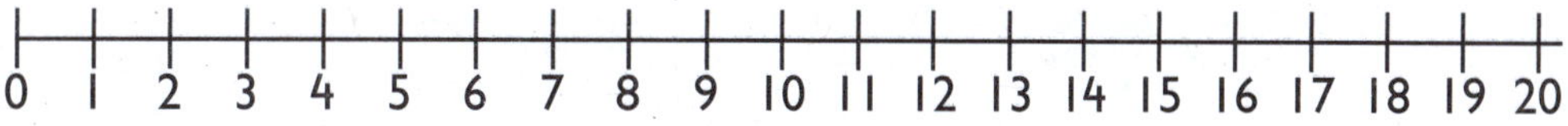

19 − 11 = ☐

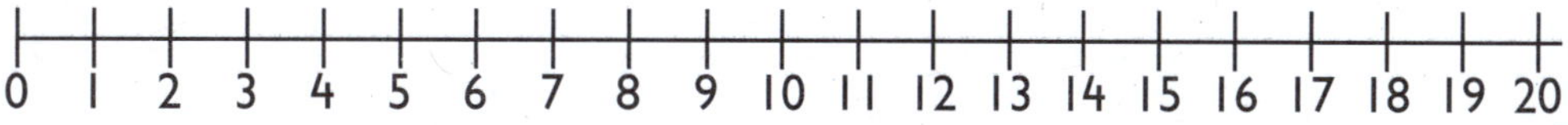

4 Work out the difference between each pair of numbers.

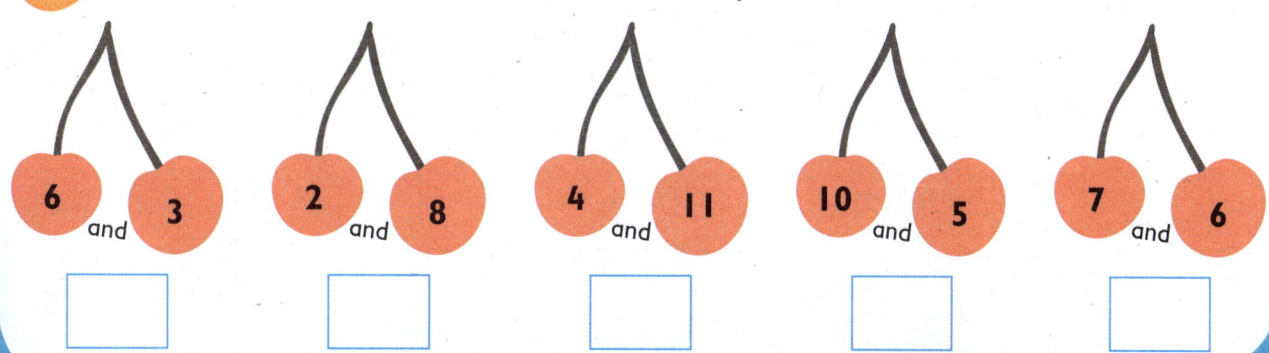

At this stage, your child is learning that subtraction involves taking away objects and counting how many are left. They are also beginning to see subtraction as finding the difference between two numbers.

Addition facts to 10

1 Draw lines to match each hand with two rings.

2 Fill in the missing number so that each number sentence totals the number on the hat.

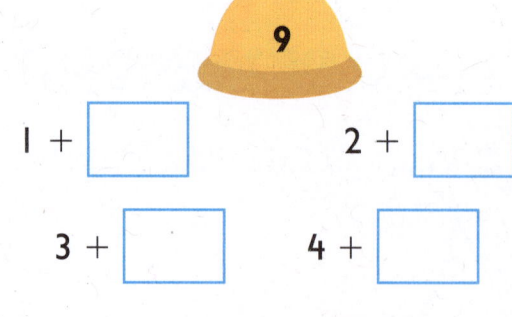

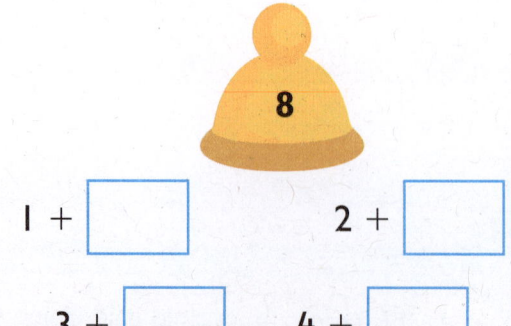

Game: Adding butterflies

You need: a 1–6 dice, some counters and some buttons

- Take turns to:
  - roll the dice and use the counters to cover that many red butterflies
  - roll the dice again and use some more counters to cover that many blue butterflies
  - count how many butterflies you have covered altogether. This is your score.

- The winner is the player with the larger score. They take a button.

- The overall winner is the first player to collect 6 buttons.

3 Fill in the missing numbers.

6 + 1 = ☐   3 + 2 = ☐   5 + 2 = ☐

7 + 3 = ☐   2 + 7 = ☐   4 + 4 = ☐

2 + ☐ = 6   1 + ☐ = 3   2 + ☐ = 10

☐ + 6 = 9   ☐ + 3 = 8   ☐ + 5 = 6

1 + ☐ = 4   ☐ + 4 = 7   ☐ + 5 = 9

6 + ☐ = 8   4 + 1 = ☐   3 + ☐ = 6

Help your child to understand that addition (unlike subtraction) can be done in any order, e.g. 2 + 5 = 5 + 2. Ask your child to count a number of objects, e.g. 5 apples. Ask them to do things such as 'Add 2', or ask 'What is 2 more?' Encourage your child to remember the larger number (say: 'Hold the larger number in your head.'), and count on to the next numbers, e.g. 5, 6, 7, so 5 add 2 is 7.

Subtraction facts to 10

1 Draw lines to match each lily pad with two frogs.

Lily pads: 4, 7, 2, 5, 6, 3

Frogs: 10 − 4, 5 − 3, 6 − 2, 7 − 0, 7 − 2, 8 − 5, 9 − 6, 7 − 1, 9 − 4, 10 − 8, 8 − 4, 10 − 3

2 Fill in the missing number so that the difference in each number sentence is the same as the number on the ladder.

10 − ☐ = 5
9 − ☐ = 5
8 − ☐ = 5
7 − ☐ = 5
6 − ☐ = 5
5 − ☐ = 5

10 − ☐ = 4
9 − ☐ = 4
8 − ☐ = 4
7 − ☐ = 4
6 − ☐ = 4
5 − ☐ = 4

10 − ☐ = 6
9 − ☐ = 6
8 − ☐ = 6
7 − ☐ = 6
6 − ☐ = 6

Game: Subtraction game

You need: a 1–6 dice and some counters in two different colours

- Take turns to roll the dice and place a counter on a matching square on the grid.
- The winner is the first player to complete a line of 4 counters. The line can be along a row or a column or a diagonal.

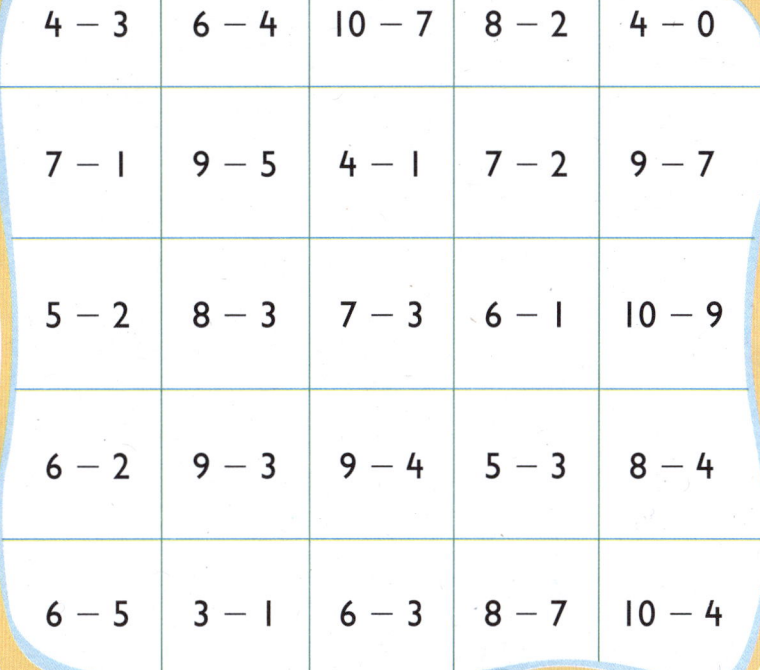

4 − 3	6 − 4	10 − 7	8 − 2	4 − 0
7 − 1	9 − 5	4 − 1	7 − 2	9 − 7
5 − 2	8 − 3	7 − 3	6 − 1	10 − 9
6 − 2	9 − 3	9 − 4	5 − 3	8 − 4
6 − 5	3 − 1	6 − 3	8 − 7	10 − 4

3 Fill in the missing numbers.

5 − 4 = ☐   8 − 5 = ☐   7 − 2 = ☐

9 − 3 = ☐   6 − 4 = ☐   10 − 7 = ☐

4 − ☐ = 2   7 − ☐ = 3   6 − ☐ = 5

☐ − 6 = 2   ☐ − 3 = 2   ☐ − 3 = 0

4 − ☐ = 3   ☐ − 5 = 4   10 − 5 = ☐

☐ − 0 = 7   8 − ☐ = 4   6 − ☐ = 3

Count out 6 counters. Hide 4 of them in a small container. Ask your child to say how many counters they can see, and how many are hidden. Repeat for other numbers to 10.

Addition facts to 20

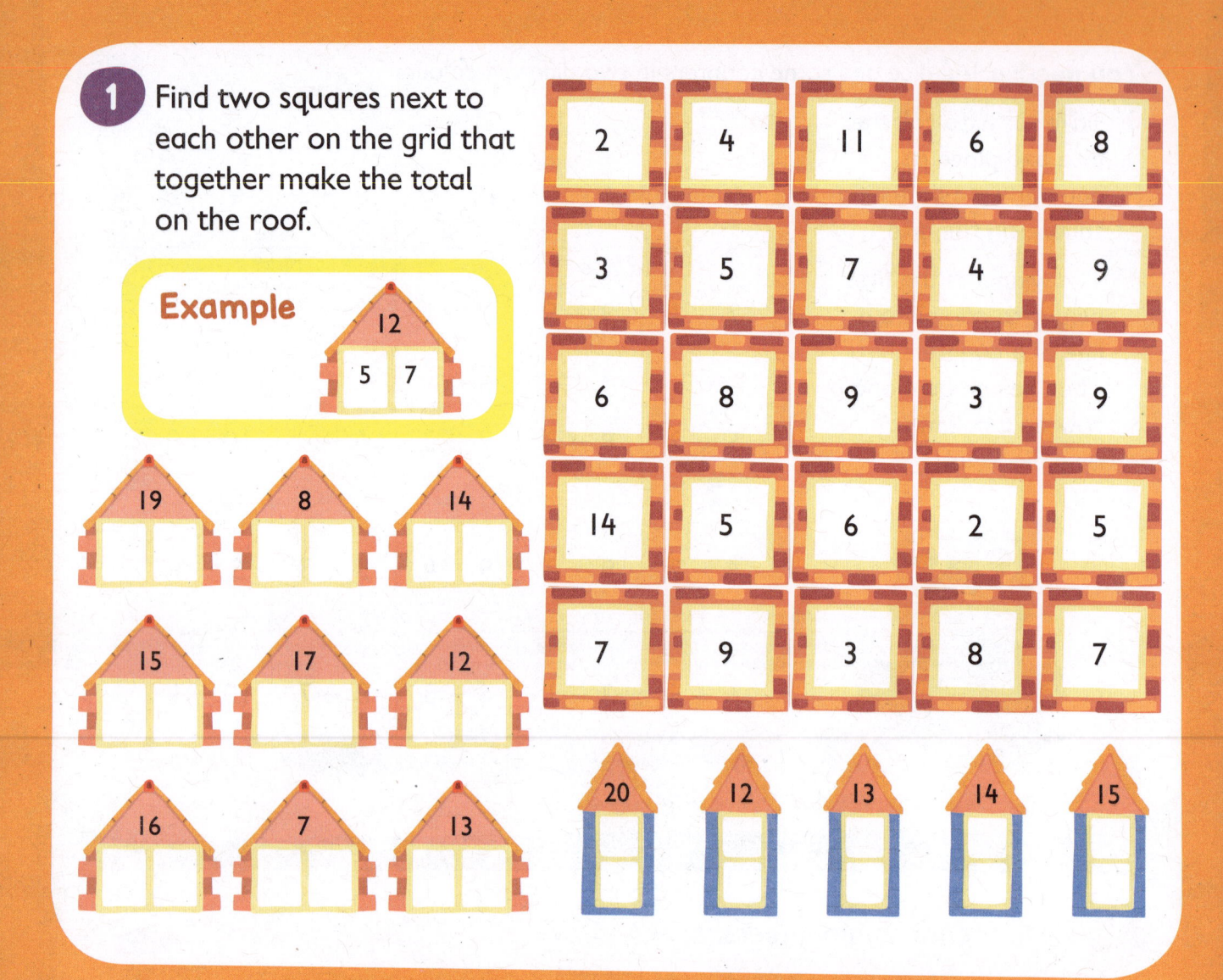

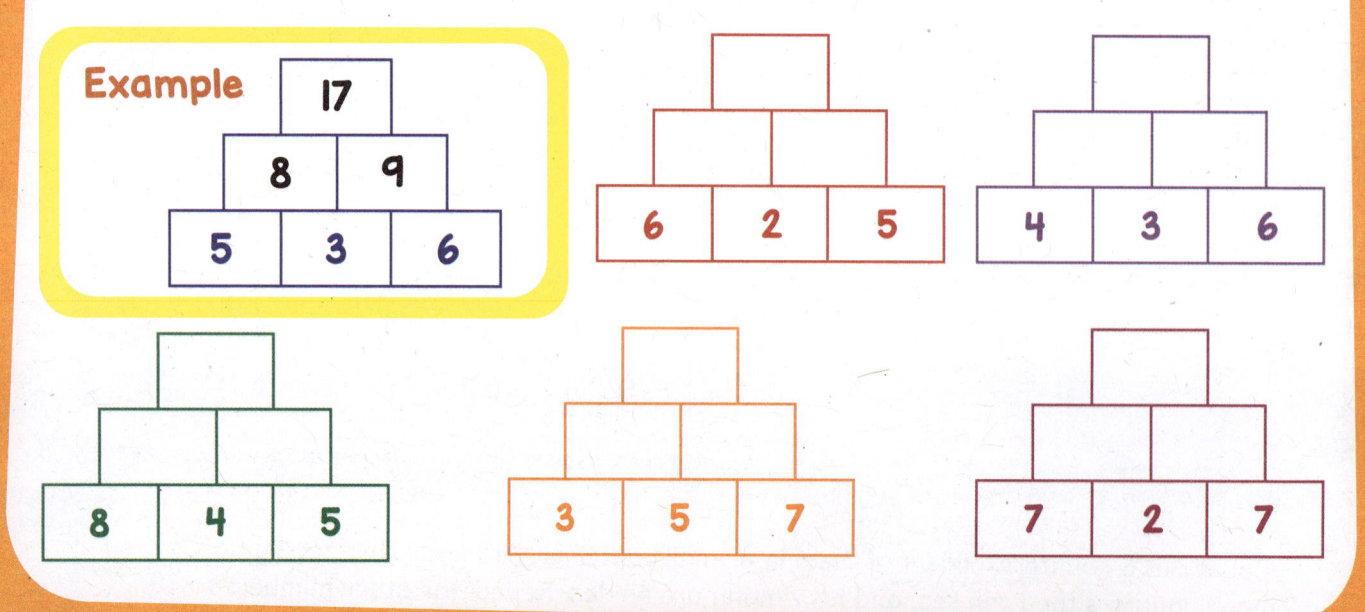

1 Find two squares next to each other on the grid that together make the total on the roof.

Example: 12 = 5 + 7

Roofs: 19, 8, 14, 15, 17, 12, 16, 7, 13

Grid:
2	4	11	6	8
3	5	7	4	9
6	8	9	3	9
14	5	6	2	5
7	9	3	8	7

Windows: 20, 12, 13, 14, 15

2 Add together pairs of numbers next to each other and write the answer in the box above.

Example:
- 17
- 8, 9
- 5, 3, 6

Pyramids:
- 6, 2, 5
- 4, 3, 6
- 8, 4, 5
- 3, 5, 7
- 7, 2, 7

Game: Choose and add the cards

You need: playing cards with the Jacks, Queens and Kings removed and 10 counters each

- Before you start:
  – decide who has the red board and who has the blue board
  – shuffle the cards and place them face down in a pile.
- Take turns to:
  – pick the top four cards, e.g.

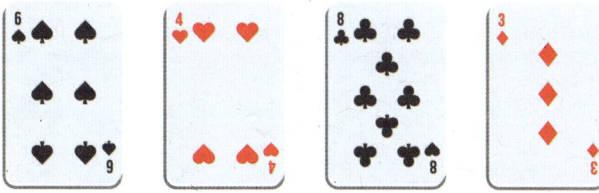

  – choose any two of the numbers and add them together, e.g. 6 + 3 = 9
  – place one of your counters on the answer on your board.
- After each player has had a turn, collect up all the cards and shuffle them again.
- The winner is the first player to cover 10 of their numbers.

Red board:
2	
3	4
5	6
7	8
9	10
11	12
13	14
15	16
17	18
19	20

Blue board:
2	
3	4
5	6
7	8
9	10
11	12
13	14
15	16
17	18
19	20

3 Complete the addition table.

+	5	8	4	7	2	3	6
5	10						
9							
12				14			
6							

Using the playing cards with the Jacks, Queens and Kings removed, shuffle and deal the cards. Players place their cards face down in a pile. Each player turns over the top card from their pile. Both players add the two numbers together. The first player to call out the correct answer wins that round and collects a counter. Play 10 rounds. Who wins more counters?

13

Subtraction facts to 20

1 Find two squares next to each other on the grid that have a difference equal to the number on the roof.

Example: roof 5, squares 3 and 8

Roofs: 3, 13, 5, 1, 8, 4, 6, 7, 9

Grid:
2	5	4	3	8
9	1	7	6	2
2	6	4	5	18
3	4	2	9	0
16	8	7	1	7

Tall houses roofs: 3, 5, 8, 4, 16

2 Find the difference between pairs of numbers next to each other and write the answer in the box above.

Example:
- Top: 3
- Middle: 4, 1
- Bottom: 6, 2, 3

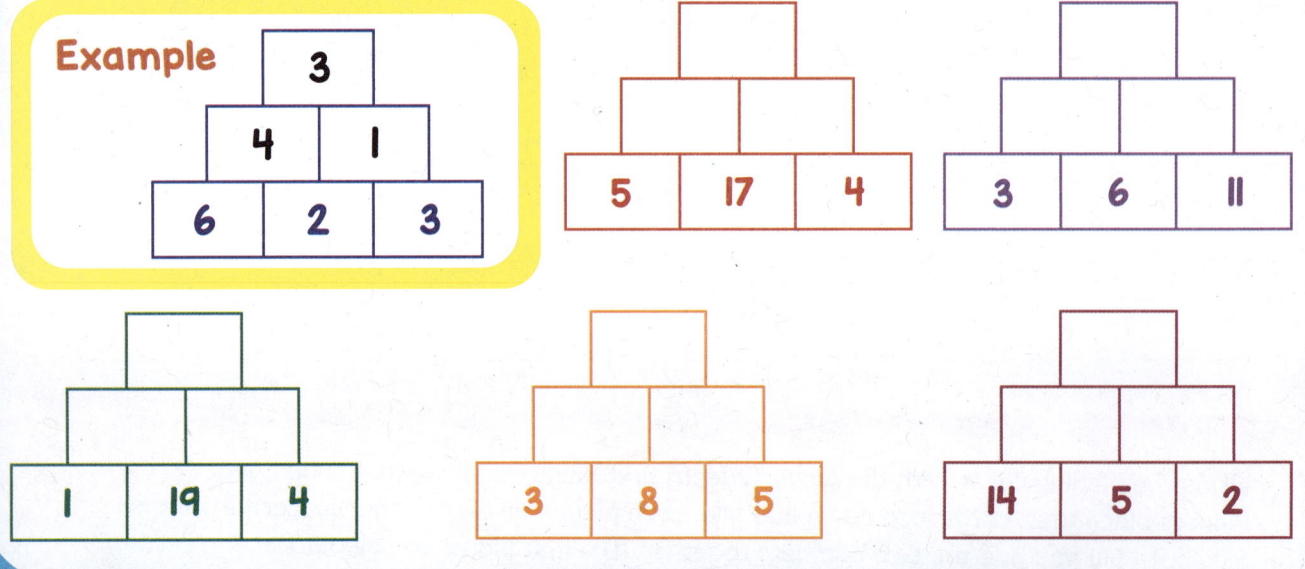

Game: Subtract the cards

You need: playing cards with all of the Kings, Queens, Jacks and three of the 10s removed and 9 counters

- Before you start:
  - decide who has the red board and who has the blue board
  - place one 10 card face up on the table
  - shuffle the remaining cards and place them face down in a pile.
- Take turns to:
  - pick the top two cards
  - place the first card on top of the 10 card to make a teen number and place the second card beside it

2	3
6	4
7	5
9	8
10	10
12	11
14	13
15	16
18	17

  - subtract the second card from the teen number, e.g. $18 - 3 = 15$

18 subtract 3 is 15

  - if the answer is on your board, cover it with a counter. If the answer isn't on your board, miss this turn.
- After each player has had a turn, collect the cards and shuffle them again.
- The winner is the first player to cover 5 of their numbers.

3 Complete the subtraction tables.

−	2	4	3	5
16				
5			1	
18				
7				2

−	5	6	3	1
7				
19	14			
6			3	
8				

Multiples of 10

Multiples of 10 are 10, 20, 30, 40, 50, …
You can use your addition number facts to work out the answers when adding together these multiples.

2 + 4 = 6
So
20 + 40 = 60

1 Fill in the missing numbers.

6 + 4 = ☐

5 + 3 = ☐

50 + 30 = ☐

6 + 3 = ☐   60 + 40 = ☐   5 + 2 = ☐

60 + 30 = ☐   50 + 20 = ☐   3 + 2 = ☐

3 + 4 = ☐

30 + 20 = ☐

4 + 4 = ☐

30 + 40 = ☐

40 + 40 = ☐

2 + 7 = ☐

1 + 8 = ☐

20 + 70 = ☐

10 + 80 = ☐

2 Fill in the missing numbers. Then draw lines to join each pair of related facts.

2 + 6 = ☐   30 + 30 = ☐   2 + 5 = ☐

50 + 40 = ☐

4 + 3 = ☐

5 + 4 = ☐

20 + 60 = ☐   1 + 7 = ☐

3 + 3 = ☐

10 + 70 = ☐   20 + 50 = ☐   40 + 30 = ☐

You can use your subtraction number facts to work out the answers when subtracting multiples of 10.

$7 - 5 = 2$
So
$70 - 50 = 20$

3 Fill in the missing numbers.

$10 - 8 = \square$

$6 - 4 = \square$ $\quad$ $5 - 2 = \square$ $\quad$ $100 - 80 = \square$

$60 - 40 = \square$ $\quad$ $50 - 20 = \square$

$4 - 1 = \square$

$7 - 3 = \square$ $\quad$ $9 - 7 = \square$ $\quad$ $40 - 10 = \square$

$70 - 30 = \square$ $\quad$ $90 - 70 = \square$

$6 - 3 = \square$

$8 - 6 = \square$ $\quad$ $3 - 2 = \square$ $\quad$ $60 - 30 = \square$

$80 - 60 = \square$ $\quad$ $30 - 20 = \square$

4 Fill in the missing numbers. Then draw lines to join each pair of related facts.

$8 - 3 = \square$ $\quad$ $5 - 3 = \square$ $\quad$ $40 - 20 = \square$

$30 - 10 = \square$

$4 - 2 = \square$ $\quad$ $7 - 6 = \square$

$3 - 1 = \square$

$80 - 30 = \square$

$70 - 60 = \square$ $\quad$ $50 - 30 = \square$

Use 1p and 10p coins to show the link between the addition and subtraction of number facts to 10 and adding and subtracting multiples of 10. For example, give your child three 1p coins then another four 1p coins and ask: 'How much money is this altogether?'. Now repeat for 10p coins.

Adding three 1-digit numbers

1 Find the totals.

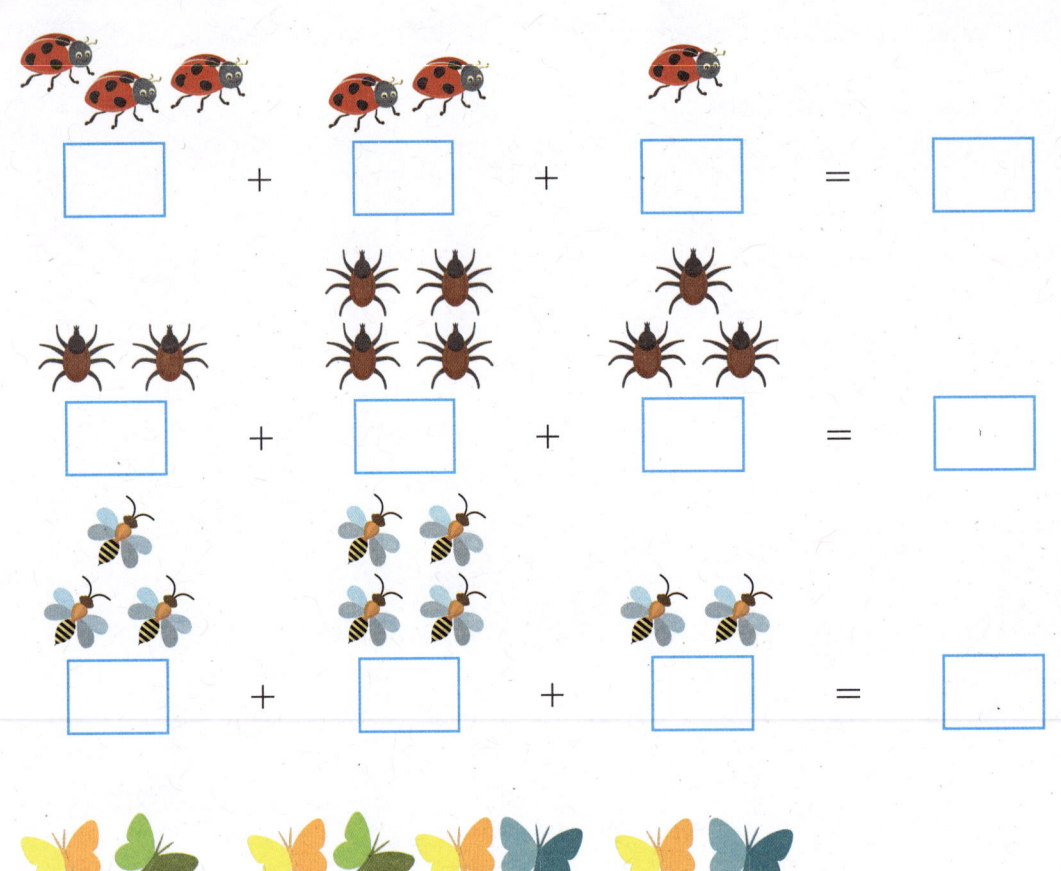

2 Find the totals in the spider's web.

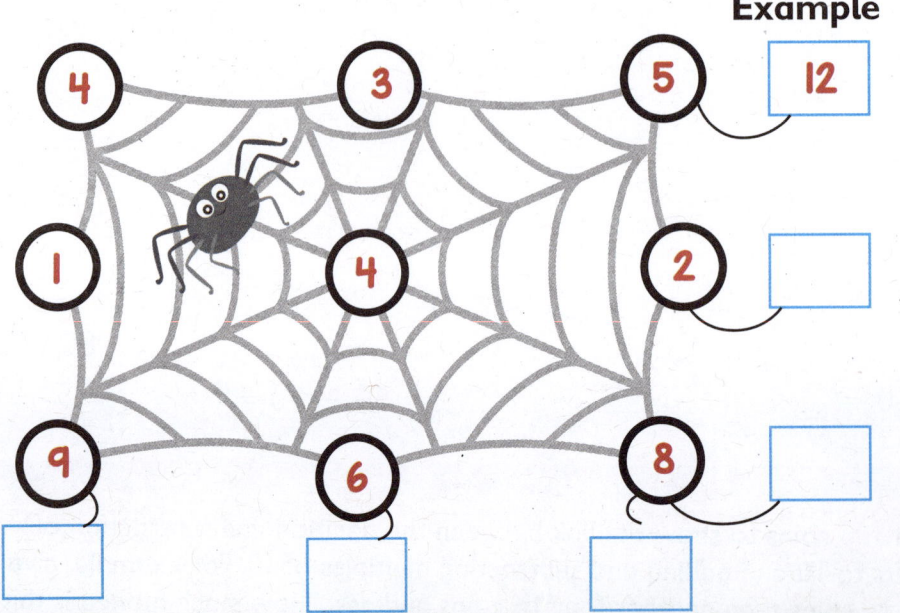

Example: 12

Addition can be done in any order. A number line helps with addition when adding more than two numbers. Count on from the **larger number**.

2 + 4 + 3 = 9

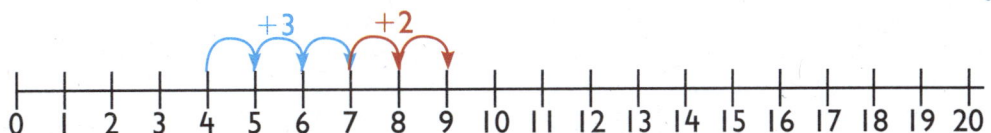

3 Use the number lines to find the totals. Start with the larger number.

6 + 2 + 1 = ☐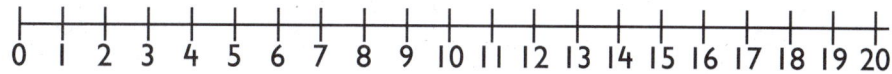

3 + 5 + 2 = ☐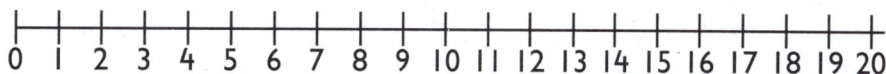

4 + 3 + 7 = ☐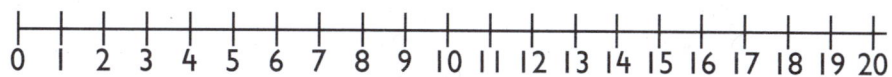

3 + 6 + 5 = ☐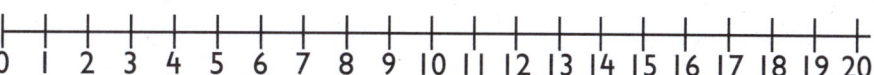

2 + 8 + 3 = ☐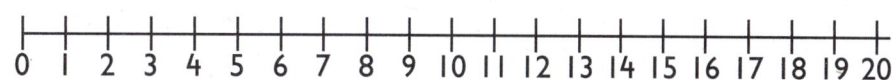

4 Complete each number sentence.

1 + 3 + 2 = ☐   2 + 4 + 3 = ☐

6 + 7 + 3 = ☐   3 + 4 + 5 = ☐

3 + 9 + 4 = ☐   5 + 7 + 8 = ☐

3 + 4 + 1 = ☐   5 + 3 + 4 = ☐

9 + 8 + 6 = ☐   6 + 2 + 3 = ☐

Your child needs to realise that more than two numbers can be added together, e.g. 3 + 6 + 4. It is important that they understand that it is more effective and efficient to put the larger number first and then add the remaining numbers.

19

Adding a 2-digit number and ones

A 1–100 number square can help you add numbers. Start with the larger number and **count on**.

Example
32 + 7 = 39

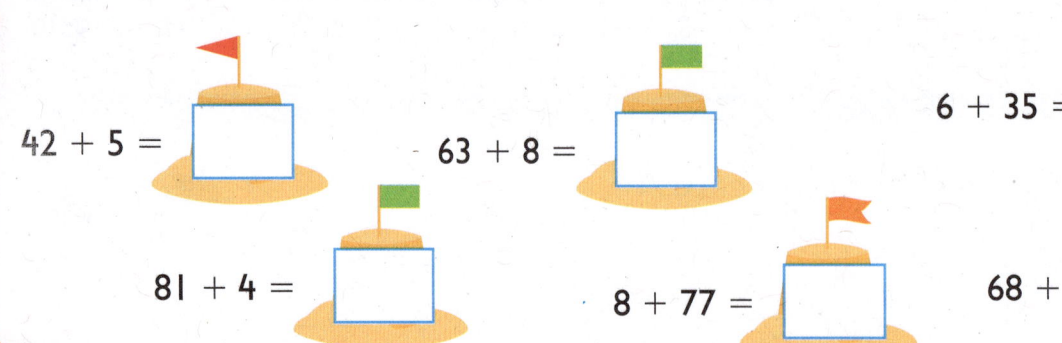

1 Use the 1–100 number square to help you add these numbers.

42 + 5 =

63 + 8 =

6 + 35 =

81 + 4 =

8 + 77 =

68 + 5 =

2 Add the number on the bucket to each of the numbers in the grid.

15

3	5	8
18		
6	4	9

24

2	8	3
5	9	7

47

3	6	9
4	7	2

Game: Adding numbers

You need: 2 paperclips, 2 pencils and some counters in two different colours

- Take turns to:
  - spin both spinners, e.g. 5 and 18
  - add the two numbers together, i.e. 23
  - place a counter on the matching flag.

- The winner is the first player to complete a line of 3 counters. The line can be along a row or a column or a diagonal.

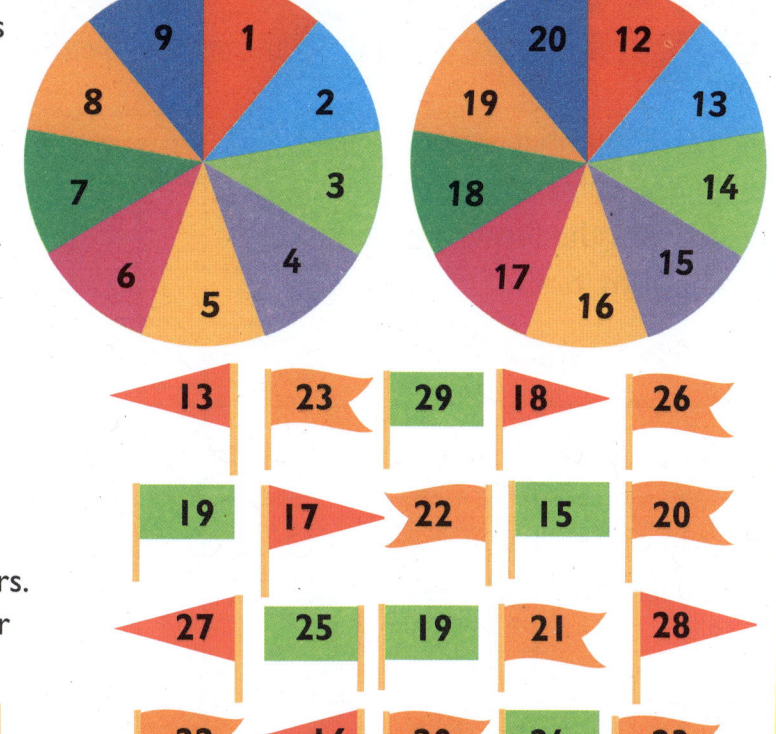

3. Draw lines to match a bucket with any spade. Then, on the spade, write the total of the two numbers.

Roll a 1–6 dice twice, e.g. 3 and 5, and make a 2-digit number, i.e. 35 or 53. Ask your child to add a 1-digit number to the 2-digit number, e.g. 35 + 7.

Subtracting a 2-digit number and ones

A 1–100 number square can help you subtract numbers. Start with the larger number and **count back**.

Example
47 − 6 = 41

1	2	3	4	5	6	7	8	9	10
11	12	13	14	15	16	17	18	19	20
21	22	23	24	25	26	27	28	29	30
31	32	33	34	35	36	37	38	39	40
41	42	43	44	45	46	47	48	49	50
51	52	53	54	55	56	57	58	59	60
61	62	63	64	65	66	67	68	69	70
71	72	73	74	75	76	77	78	79	80
81	82	83	84	85	86	87	88	89	90
91	92	93	94	95	96	97	98	99	100

1 Use the 1–100 number square to help you subtract these numbers.

93 − 8 =

52 − 7 =

81 − 5 =

60 − 7 =

35 − 4 =

74 − 3 =

2 Find the difference between the number on the car and the numbers in the grid.

25

6	3	7
19		
4	9	5

38

2	5	9
3	8	6

43

4	7	5
1	9	8

Game: Subtracting numbers

- Take turns to:
  - spin both spinners, e.g. 4 and 13
  - take away the smaller number from the larger number, i.e. 9
  - place a counter on the matching sail.

- The winner is the first player to complete a line of 4 counters. The line can be along a row or a column or a diagonal.

You need: 2 paperclips, 2 pencils and some counters in two different colours

3. Draw lines to match a balloon with any basket. Then, on the basket, write the difference between the two numbers.

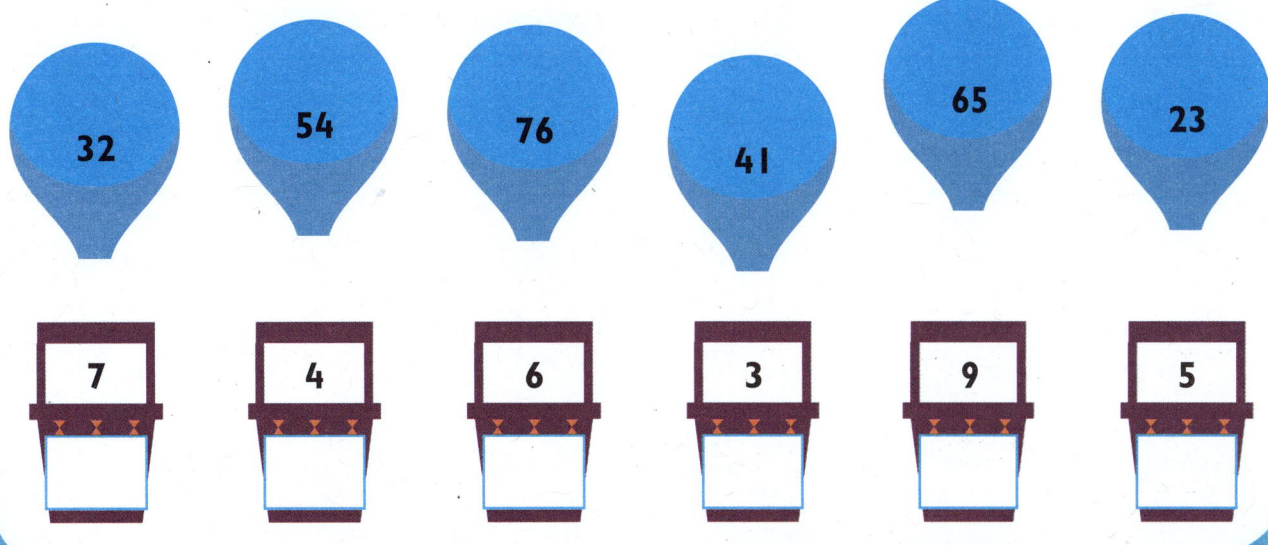

Roll a 1–6 dice twice, e.g. 6 and 2, and make a 2-digit number, i.e. 62 or 26. Ask your child to subtract a 1-digit number from the 2-digit number, e.g. 62 − 5.

Adding a 2-digit number and tens

We can use an empty number line to add a tens number to a 2-digit number.

76 + 40 = 116

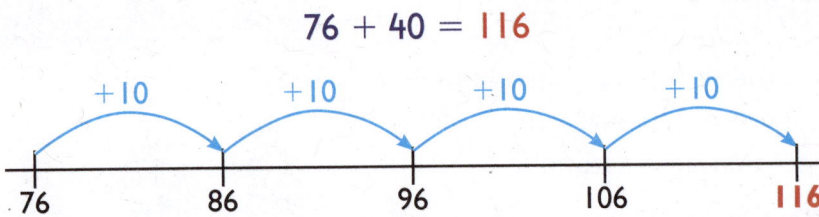

1 Use an empty number line to help you add these numbers.

45 + 50 = ☐

52 + 30 = ☐

27 + 70 = ☐

38 + 40 = ☐

2 Work out the total of each pair of number cards.

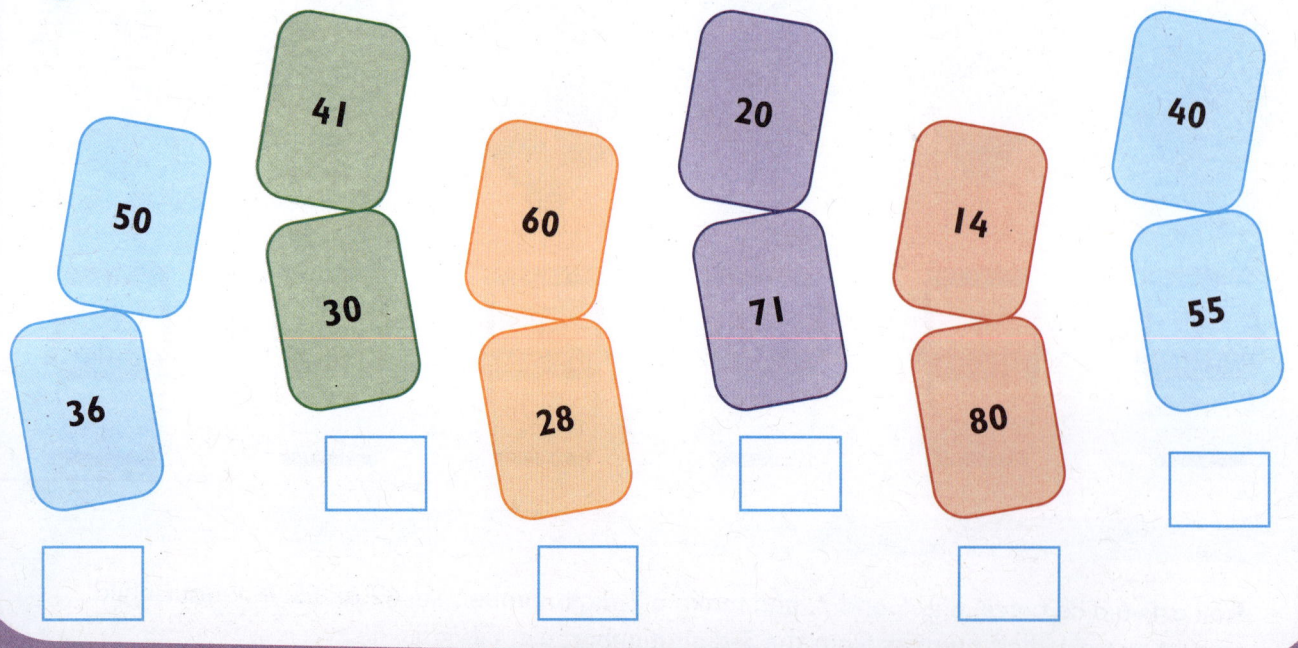

Game: Adding tens

- Before you start:
  - cover each number on the bears with a counter.

- Take turns to:
  - spin the spinner
  - remove a counter from one of the bears
  - add the number on the bear to the spinner number.

- The player with the larger total takes both counters.

- The winner is the first player to collect 6 counters.

You need: 10 counters, pencil and paperclip

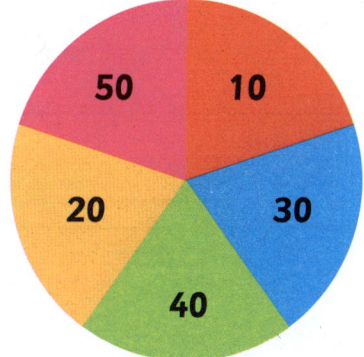

③ Draw a line between each pair of balloons that together total a number on one of the boxes.

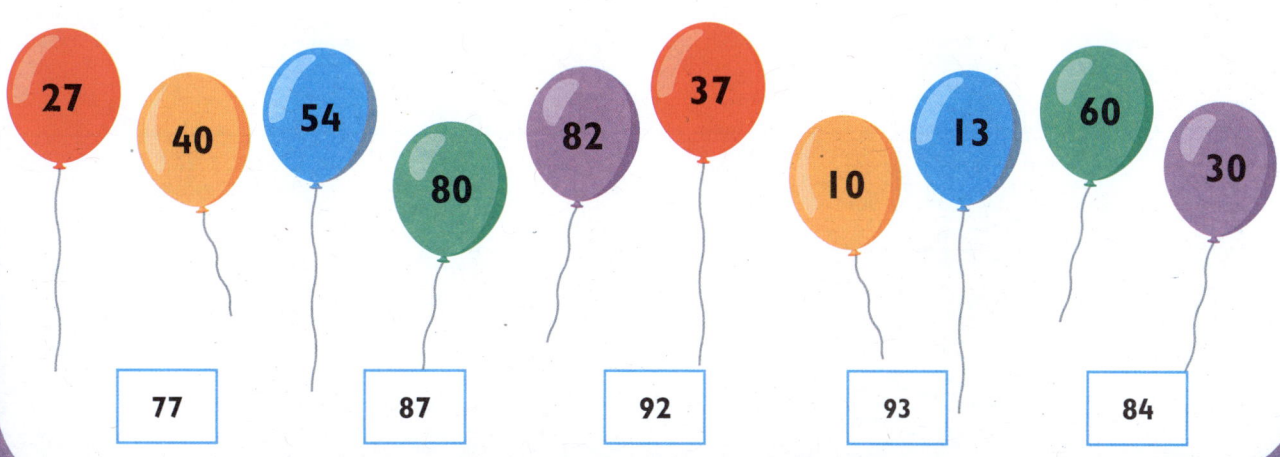

Write down a 2-digit number and ask your child to add different tens numbers to it. Ensure that they realise that the ones digit does not change. Do this with several different 2-digit numbers.

25

Subtracting a 2-digit number and tens

We can use an empty number line to subtract a tens number from a 2-digit number. You can work out the answer to a subtraction calculation by:
- taking away (counting back from the larger number) or
- finding the difference (counting on from the smaller number).

$$52 - 30 = 22$$

Taking away

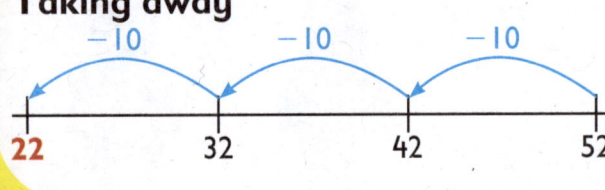

Finding the difference

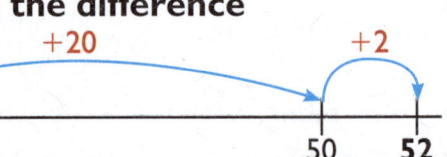

1 Use an empty number line to answer these calculations.

67 − 40 =

85 − 30 =

54 − 20 =

73 − 50 =

2 Work out the difference between each pair of number cards.

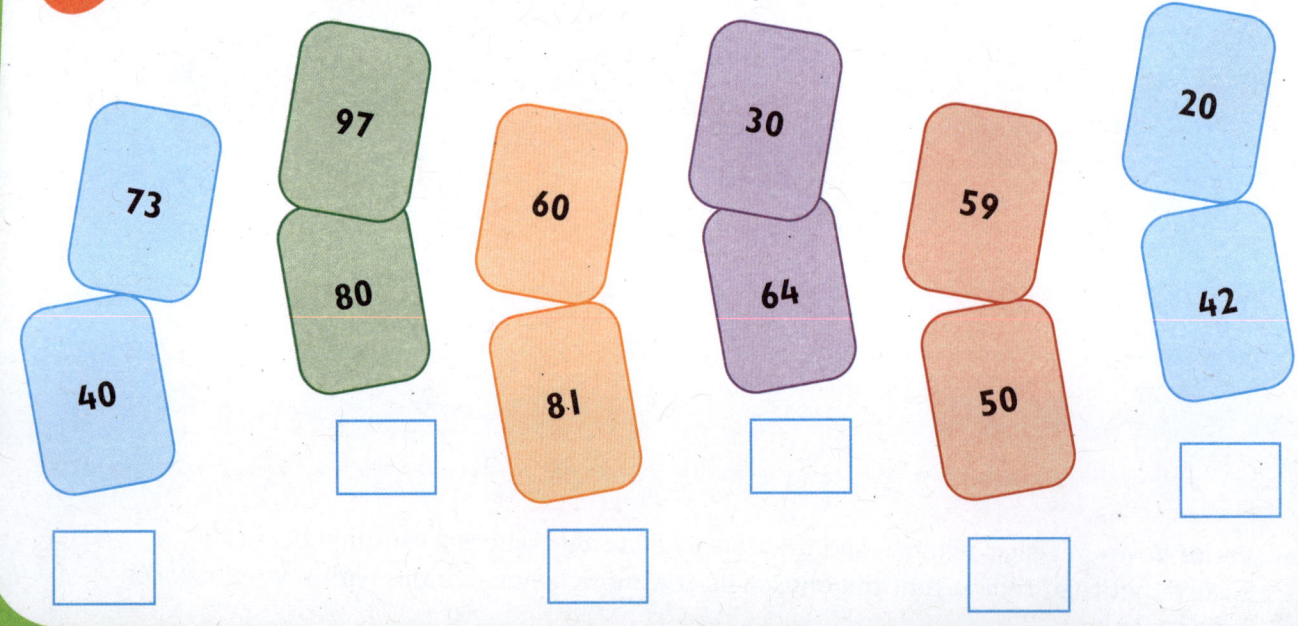

Game: Subtracting tens

You need: 10 counters, pencil and paperclip

- Before you start:
  – cover each number on the umbrellas with a counter.

- Take turns to:
  – spin the spinner
  – remove a counter from one of the umbrellas
  – subtract the spinner number from the umbrella number.

- The player with the smaller answer keeps both counters.

- The winner is the first player to collect 6 counters.

3 Complete the subtraction table.

−	40	20	50	30	60	10
77						
83						
96		76				
68						
92						

Point to one of the larger numbers on the umbrellas in the game above and ask your child to subtract different tens numbers from it. Ensure they understand that the ones digit does not change. Do this with several different numbers from the umbrellas.

Adding two 2-digit numbers

Addition can be done in any order, so 25 + 38 is the same as 38 + 25.
- Put the larger number first.
- Count on the number of tens in the smaller number.
- Count on the number of ones in the smaller number.

Look at these different methods for adding pairs of 2-digit numbers.

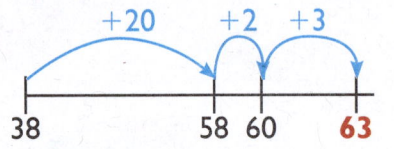 OR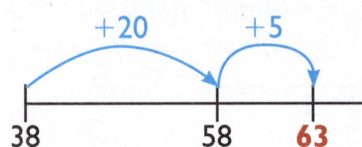

OR   OR   OR
38 + 25 = 38 + 20 + 5   38 + 25 = 30 + 20 + 8 + 5   30 + 8
       = 58 + 5   = 50 + 13   + 20 + 5
       = 63   = 63   50 + 13 = 63

1 Work out the answers. Write down your thinking in the thought bubbles.

27 + 42 = ☐   45 + 28 = ☐

56 + 39 = ☐   37 + 35 = ☐

33 + 48 = ☐   42 + 57 = ☐

28

Game: Spin and roll

- Take turns to:
  - spin the spinner
  - roll the dice and use the numbers to make a 2-digit number
  - add the spinner number to the 2-digit dice number.

- The player with the larger total wins the round and takes a counter.

- The winner is the first player to collect 5 counters.

You need: 10 counters, two 1–6 dice, pencil and paperclip

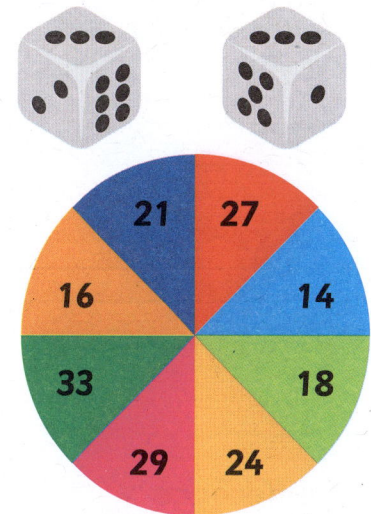

2 Draw lines to join each addition problem with its answer.

| 26 + 34 | 52 + 79 | 68 + 53 | 35 + 47 | 76 + 19 | 44 + 28 |

| 82 | 72 | 131 | 60 | 121 | 95 |

3 Fill in the missing numbers.

33 + 25 = ☐   29 + ☐ = 91   ☐ + 19 = 52

46 + 31 = ☐   53 + 27 = ☐   46 + ☐ = 94

23 + ☐ = 72   79 + ☐ = 97   65 + 28 = ☐

☐ + 14 = 79   ☐ + 47 = 73   39 + ☐ = 86

Your child is beginning to add pairs of 2-digit numbers mentally. If necessary, assist them in making jottings to help this process.

Subtracting two 2-digit numbers

Look at these different methods for subtracting pairs of 2-digit numbers.

$$73 - 45 = 28$$

Taking away

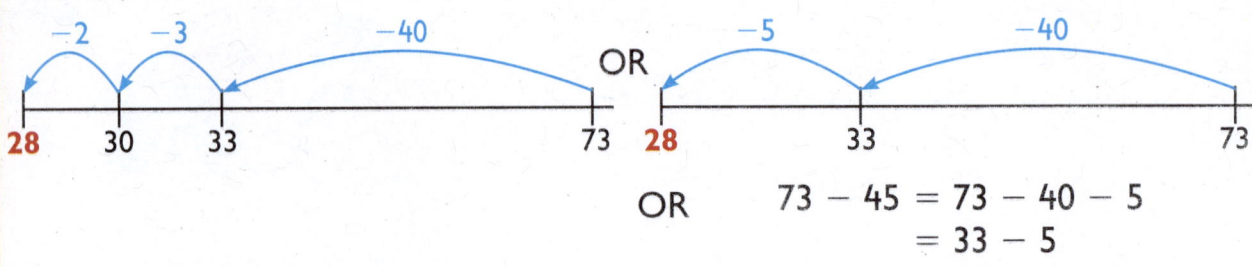

OR

$$73 - 45 = 73 - 40 - 5$$
$$= 33 - 5$$
$$= 28$$

Find the difference

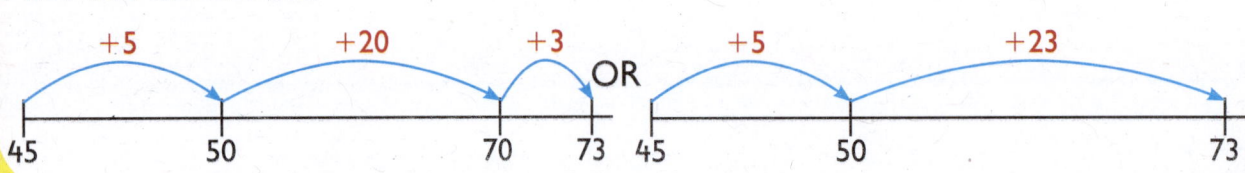

1 Work out the answers. Write down your thinking in the thought bubbles.

63 − 21 = ☐   77 − 34 = ☐

82 − 56 = ☐   54 − 28 = ☐

75 − 47 = ☐   93 − 36 = ☐

30

Game: Deal the digits

You need: a pack of playing cards with the 10s, Jacks, Queens and Kings removed, 10 counters

- Shuffle the cards and place them face down in a pile.
- Take turns to:
  - take 4 cards from the top of the pile
  - lay them face up on the table to make two 2-digit numbers
  - find the difference between the two numbers.

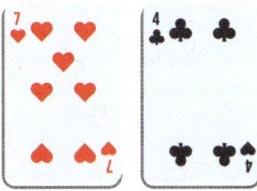 

74 minus 51 equals 23

- The player with the smaller difference between their two numbers wins the round and takes a counter.
- The winner is the first player to collect 5 counters.

2 Draw lines to join each subtraction problem with its answer.

| 38 − 24 | 71 − 43 | 85 − 59 | 47 − 22 | 58 − 35 | 62 − 58 |

| 25 | 14 | 28 | 4 | 23 | 26 |

3 Fill in the missing numbers.

74 − 22 = ☐   ☐ − 35 = 14   ☐ − 38 = 49

67 − ☐ = 31   61 − ☐ = 36   53 − ☐ = 16

43 − 18 = ☐   95 − ☐ = 23   69 − 17 = ☐

84 − 51 = ☐   46 − 29 = ☐   92 − ☐ = 28

Your child is beginning to subtract pairs of 2-digit numbers mentally. If necessary, assist them in making jottings to help this process.

Answers

Understanding addition

Page 4

1. 5 + 4 = 9   3 + 7 = 10
   2 + 6 = 8   2 + 4 = 6
   8 + 5 = 13 9 + 7 = 16

2. 3 + 5 = 8
   2 + 4 = 6
   4 + 2 = 6
   5 + 6 = 11

Page 5

3. 4 + 6 = 10   8 + 3 = 11
   5 + 7 = 12   6 + 9 = 15
   12 + 5 = 17

4.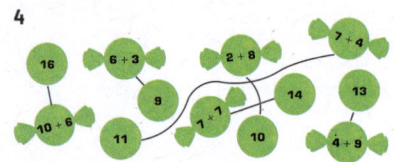

Understanding subtraction

Page 6

1. 5 − 2 = 3   6 − 4 = 2
   9 − 5 = 4   8 − 3 = 5
   12 − 6 = 6 10 − 2 = 8

2. 7 − 5 = 2   10 − 6 = 4
   18 − 9 = 9 15 − 8 = 7

Page 7

3. 7 − 2 = 5   16 − 9 = 7
   12 − 5 = 7 19 − 11 = 8

4. 3 6 7 5 1

Addition facts to 10

Page 8

1. 4 = 1 + 3 / 2 + 2   5 = 4 + 1 / 3 + 2
   6 = 3 + 3 / 2 + 4   7 = 5 + 2 / 3 + 4
   8 = 6 + 2 / 4 + 4   9 = 3 + 6 / 5 + 4

2. 6 = 1 + 5 / 2 + 4 / 3 + 3
   9 = 1 + 8 / 2 + 7 / 3 + 6 / 4 + 5
   7 = 1 + 6 / 2 + 5 / 3 + 4
   8 = 1 + 7 / 2 + 6 / 3 + 5 / 4 + 4

Page 9

3. 6 + 1 = 7   3 + 2 = 5
   5 + 2 = 7   7 + 3 = 10
   2 + 7 = 9   4 + 4 = 8
   2 + 4 = 6   1 + 2 = 3
   2 + 8 = 10 3 + 6 = 9
   5 + 3 = 8   1 + 5 = 6
   1 + 3 = 4   3 + 4 = 7
   4 + 5 = 9   6 + 2 = 8
   4 + 1 = 5   3 + 3 = 6

Subtraction facts to 10

Page 10

1. 2 = 5 − 3 / 10 − 8   3 = 8 − 5 / 9 − 6
   4 = 6 − 2 / 8 − 4   5 = 9 − 4 / 7 − 2
   6 = 10 − 4 / 7 − 1   7 = 7 − 0 / 10 − 3

2. 5   4   6
   10 − 5   10 − 6   10 − 4
   9 − 4   9 − 5   9 − 3
   8 − 3   8 − 4   8 − 2
   7 − 2   7 − 3   7 − 1
   6 − 1   6 − 2   6 − 0
   5 − 0   5 − 1

Page 11

3. 5 − 4 = 1   8 − 5 = 3
   7 − 2 = 5   9 − 3 = 6
   6 − 4 = 2   10 − 7 = 3
   4 − 2 = 2   7 − 4 = 3
   6 − 1 = 5   8 − 6 = 2
   5 − 3 = 2   3 − 3 = 0
   4 − 1 = 3   9 − 5 = 4
   10 − 5 = 5   7 − 0 = 7
   8 − 4 = 4   6 − 3 = 3

Addition facts to 20

Page 12

1.

Other answers are possible.

2.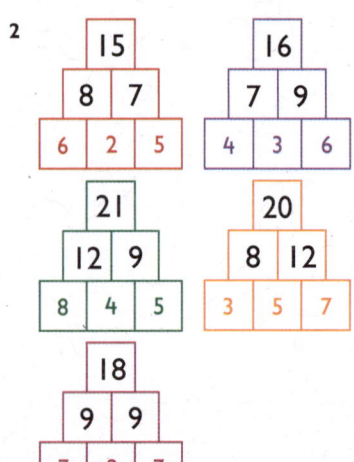

Page 13

3.

+	5	8	4	7	2	3	6
5	10	13	9	12	7	8	11
9	14	17	13	16	11	12	15
12	17	20	16	19	14	15	18
6	11	14	10	13	8	9	12

Subtraction facts to 20

Page 14

1.

Other answers are possible.

2.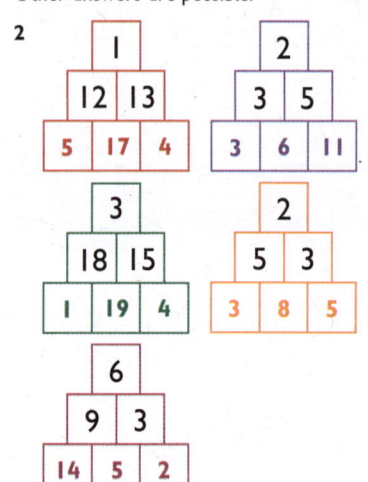